Polygons, Polyhedrons, and Other Shapes for Grades Pre-K through 5th

70 Times 7 Math

Published by 70 Times 7 Math (A division of Habakkuk Educational Materials)

Copyright © 2019-2022 by 70 Times 7 Math. All rights reserved.

POLYGONS, POLYHEDRONS, AND OTHER SHAPES FOR GRADES PRE-K THROUGH 5TH

Copyright © 2019-2022 by 70 Times 7 Math

No part of this book without a reproducible notice affixed to the footnote may be reproduced in any form or by any electronic or mechanical means, including information storage and retrieval systems, without the written consent of the publisher. If a page is specified as reproducible, the reproduction is permitted for non-commercial, classroom use only. Please address your inquiries to Habakkuk@cox.net.

ISBN (Hardback Edition): 978-1-954796-50-8
ISBN (Paperback Edition): 978-1-954796-23-2

Image on the front cover: © BlueRingMedia/Shutterstock.com

Interior illustrations: Several copyrighted images in the interior of this book are used under license from stock.adobe.com. These include the square and heart characters by yusufdemirci; the colored spheres by Alla; and the cone by anatolir on the page titled "Solid Figures that Are Not Classified as Polyhedrons."

© 70 Times 7 Math, © stock.adobe.com, and openclipart.org

Printed and bound in the United States of America

Published by 70 Times 7 Math
(A division of Habakkuk Educational Materials)

Visit www.habakkuk.net

Polygons, Polyhedrons, and Other Shapes for Grades Pre-K through 5th begins by teaching children the basic shapes, such as circles, squares, and rectangles, as well as familiar colors. They can be used as digital flashcards when opened with an eBook and are appropriate for children as young as three. Older students (those in at least kindergarten) are then taught what a polygon is and will learn "The Polygon Song," which shows how to identify a polygon by the number of sides it has and a polyhedron by the number of polygonal faces. Examples of these two- and three-dimensional shapes are provided. Students will also learn to identify other 3-dimensional shapes which are not classified as polyhedrons (i.e., cones, cylinders, and spheres).

An exam to test their comprehension over the book's content is included towards the end of the book, and the answer key is provided. The same test is also available online as a computer-based test. You can access the test free of charge by visiting the website of Habakkuk Educational Materials. (See the next page for more information.) Students can take the test as many times as they like and might originally be allowed to use their books to guide them through it. The computer will notify them if an answer is correct or incorrect, and their grades will be displayed after completing a test.

For more information or to contact Habakkuk Educational Materials, please visit the website below.

https://www.habakkuk.net/

Access to the Free Computer-Based Test

A **computer-based test** specially designed to assess a student's knowledge of the information covered in this book is available by visiting the Habakkuk Educational Materials website. Follow the two steps below.

1. Go to https://www.habakkuk.net/. From the homepage, click on "Free Computer-based Tests to Complement Books by Habakkuk Educational Materials." Then click the image with an illustration of this book to be directed to an LMS.
2. From the LMS, you should see the title of the book you are using. Click the downward arrow beneath the title (it's on the right-hand side). Then click on the test under the description, followed by enter.

After submitting an answer to a question, the computer will notify test takers if their answer is correct or incorrect. After entering their answer to the final question and clicking "Continue," a percentage grade will be available and a "Review Course" option to review any incorrect answers will also be accessible.

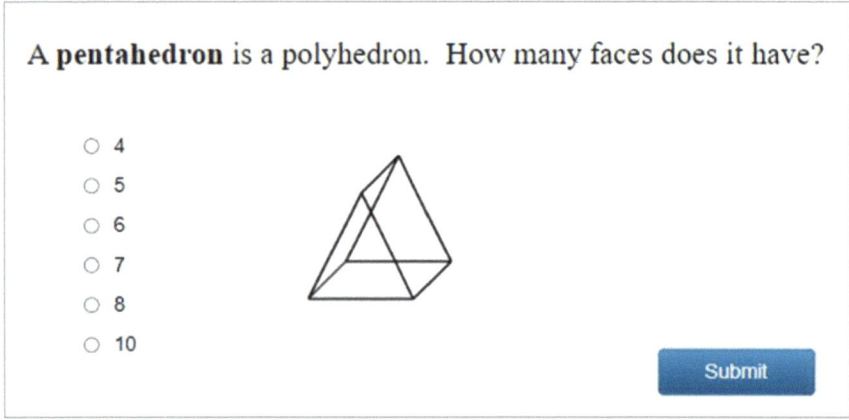

The test serves two purposes:
(1) to assess a student's mastery of the content in this book; and
(2) to be used for review games.

Using the tests for review games: The computer-based tests can be used in correspondence with various board games that have a pathway from start to finish. Students playing the game would answer one of the questions (most of which are multiple choice), and if the computer confirms that the answer is correct, the student could roll dice or spin a spinner and move his or her playing piece the corresponding number of spaces on the path. If you visit the "Free Teaching Materials" page of the Habakkuk Educational Materials website, there are free gameboards that can be printed on cardstock. (Click on "Board games and more to complement Habakkuk Educational Materials' Bible, reading, language, math, science, and social studies materials.")

Two-Dimensional (Plane) Shapes and Colors for Grades Pre-K and Up

Instructions: Read the name of each shape on the next several pages and have students repeat it. With regular practice, they will soon be able to identify the shapes without the teacher's assistance.

yellow

square

A square has four sides, and they are all the same size.

white

rectangle

Rectangle: A rectangle also has four sides, but only the opposite sides are equal. Can you point to the opposite sides?

brown

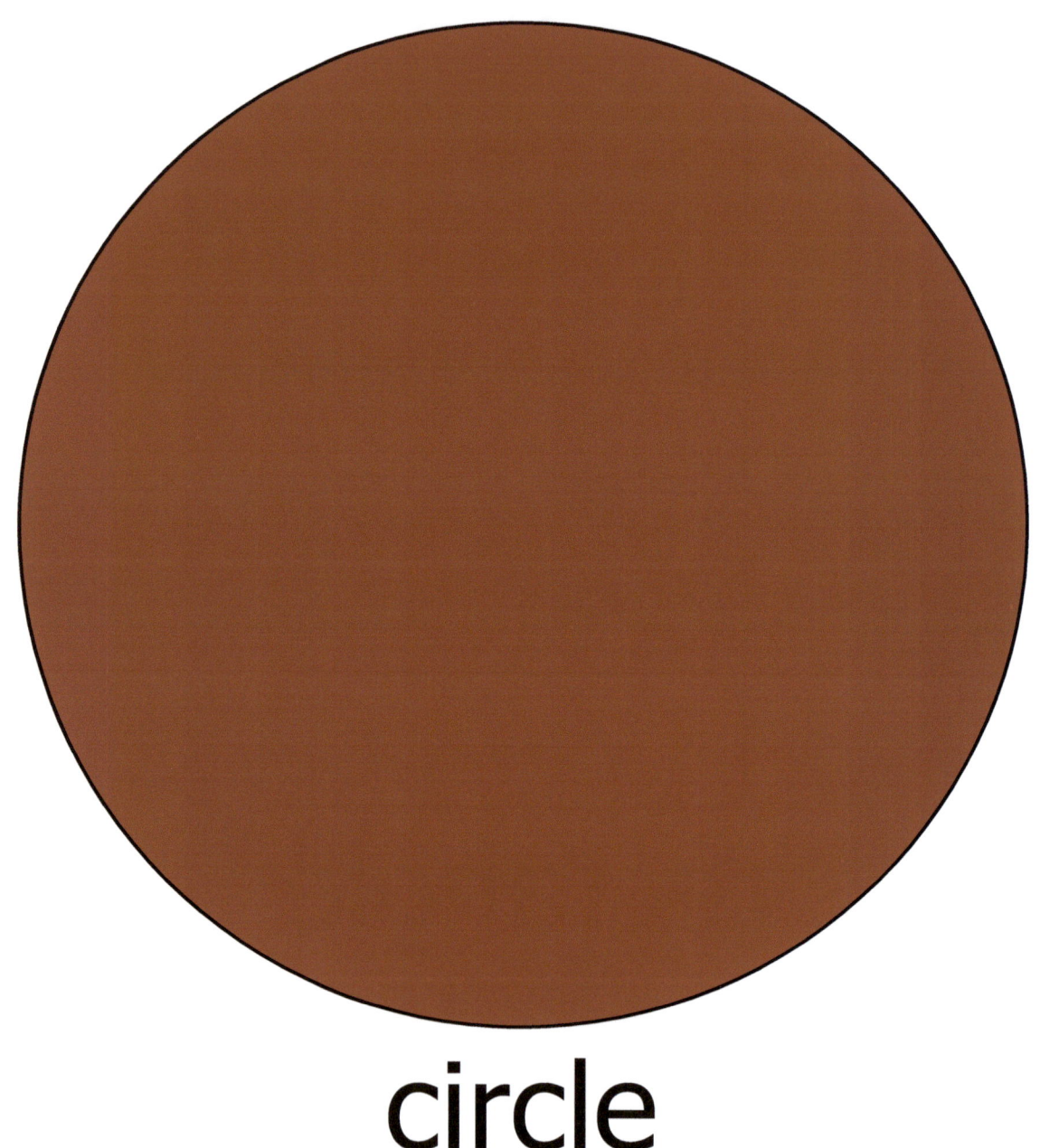

circle

Circle: A circle does not have any sides. It is completely round.

orange

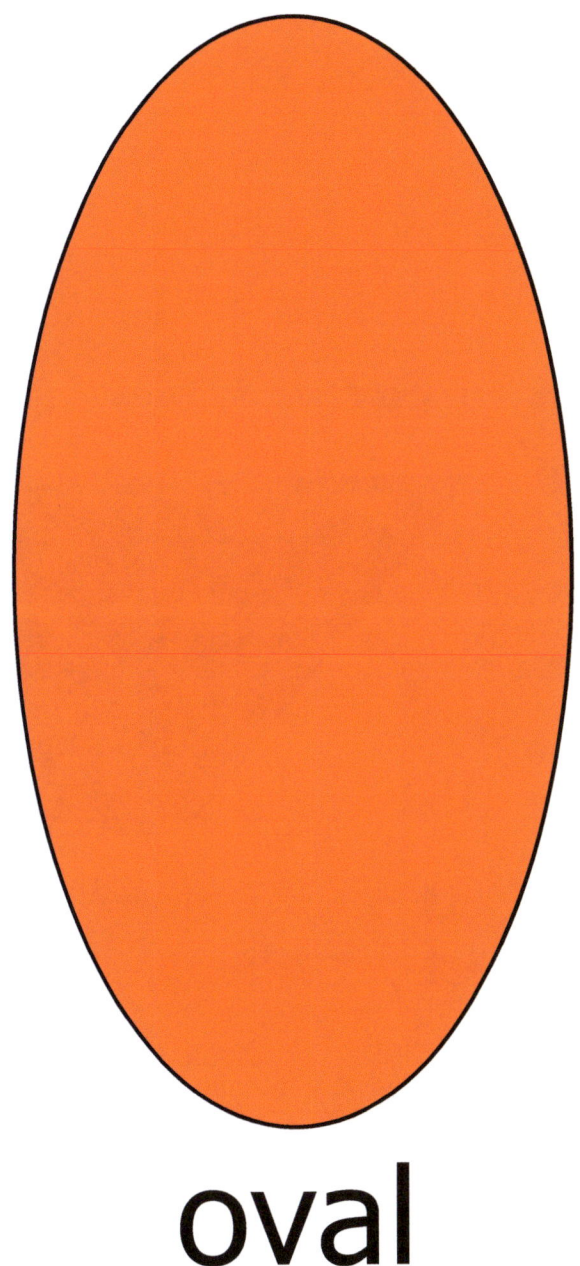

oval

Oval: An oval is also a round shape. Can you tell that it looks skinnier than the brown circle?

green

diamond

rhombus

Instruction: This shape can be referred to as either a diamond or a rhombus.

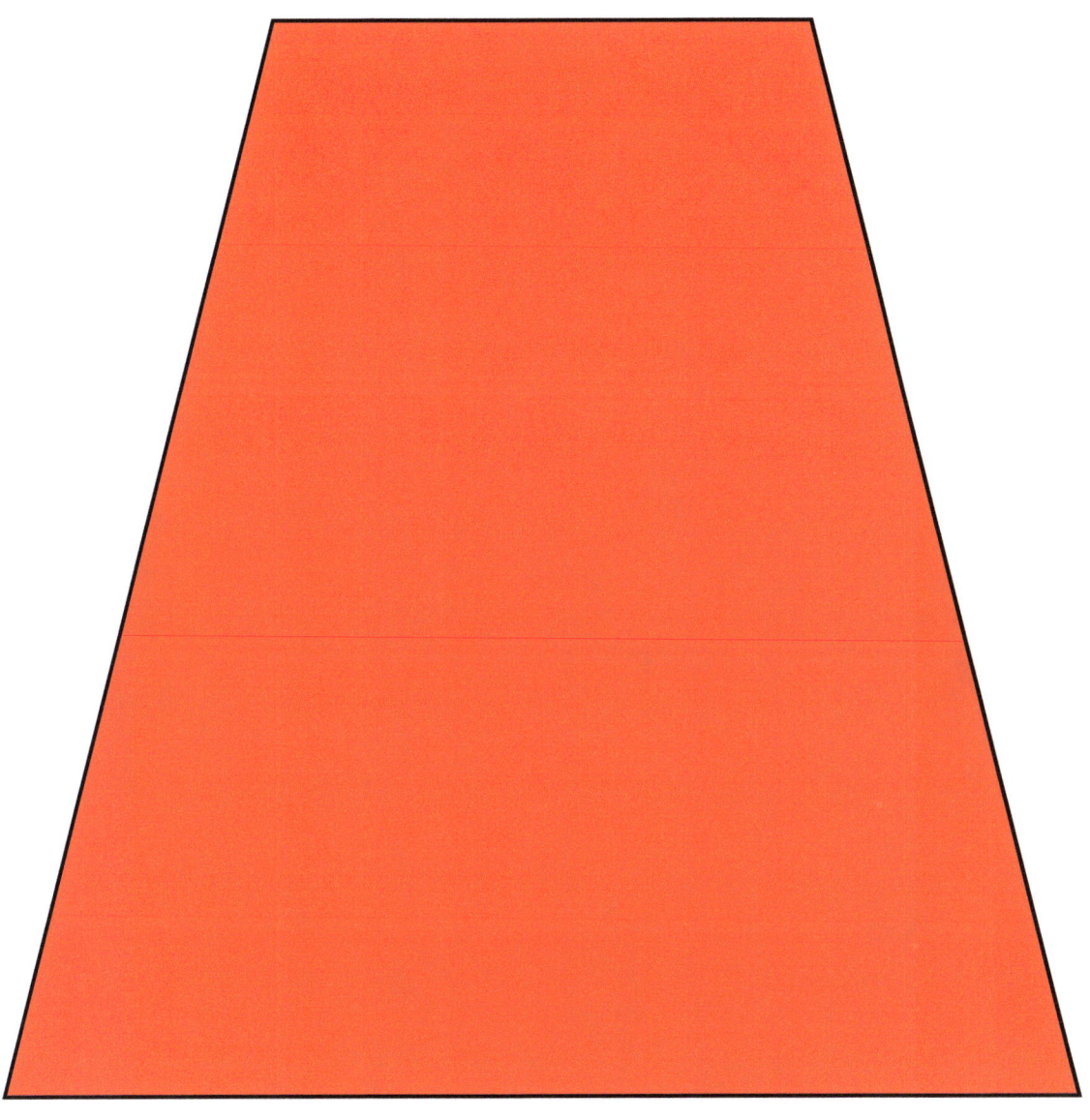

red

trapezoid

Pentagon: A pentagon has five sides. Count the sides.

gray

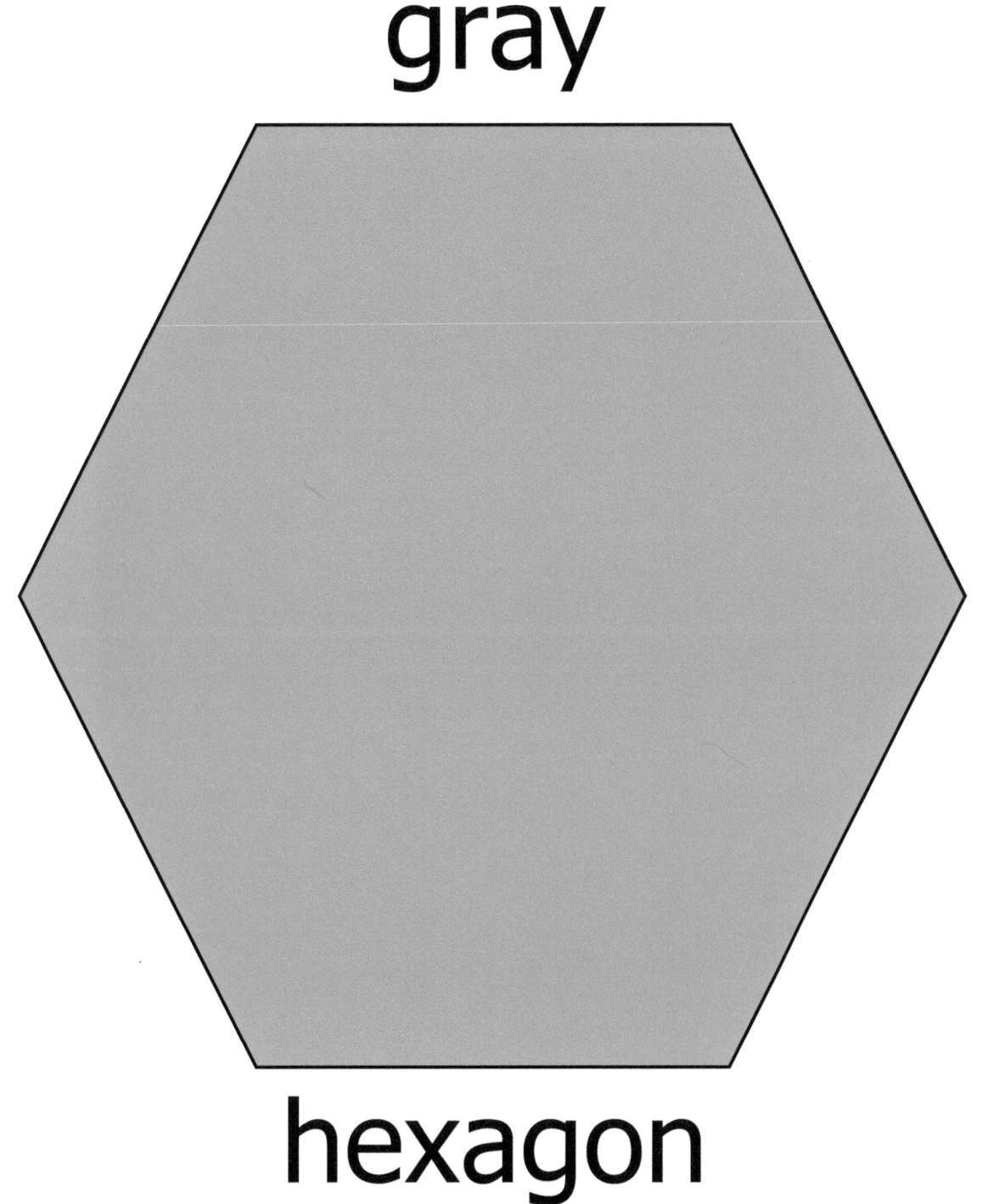

hexagon

Hexagon: A hexagon has six sides. Count the sides.

turquoise

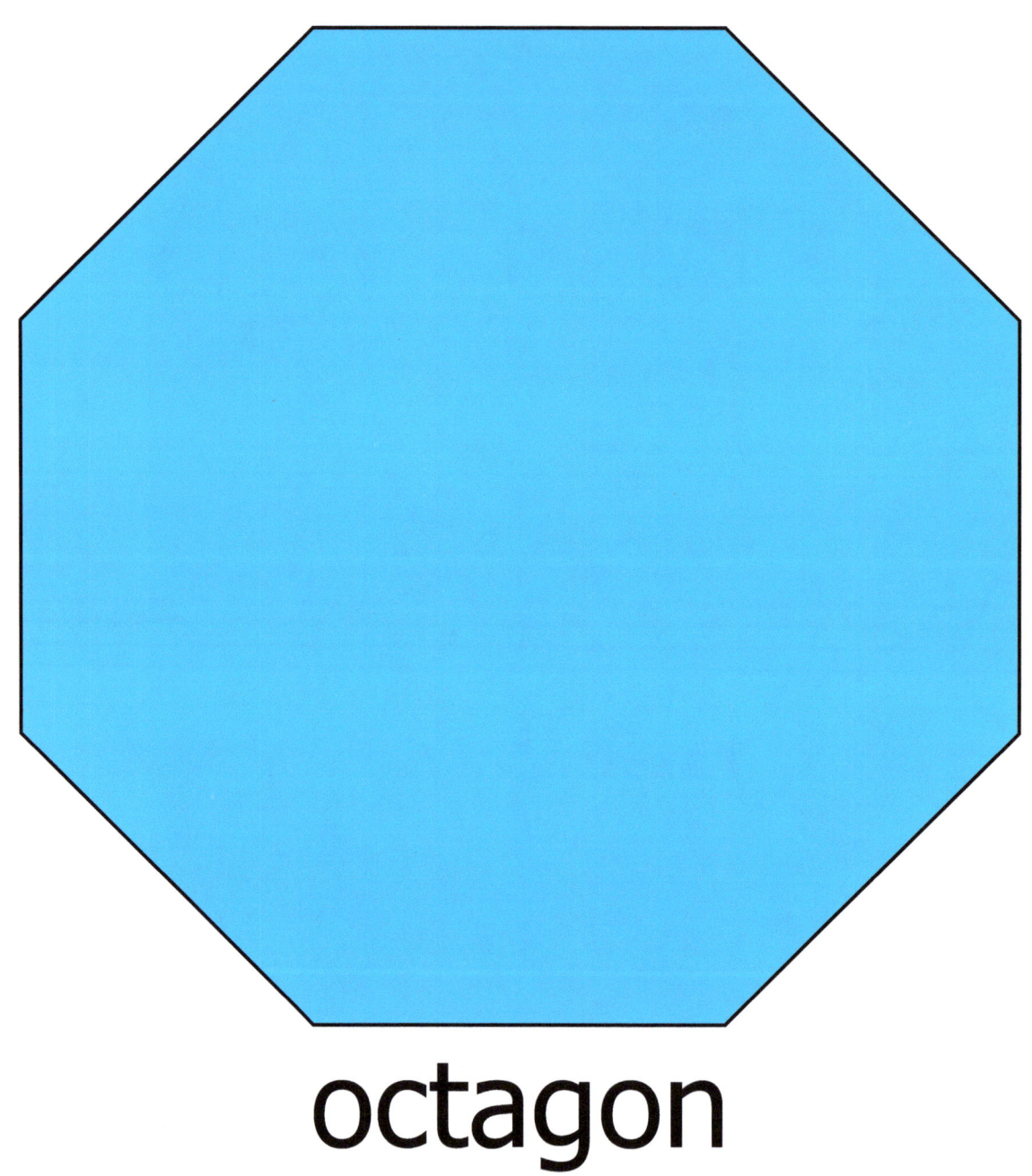

octagon

Octagon: An octagon has eight sides. Count the sides.

purple

star

blue

cross

pink

heart

You're doing great!

Triangles

Instruction: Explain to the children that although these shapes do not look exactly alike (the red one is an equilateral triangle and the yellow is a right triangle), they are still triangles because both shapes have three sides. (Count the sides with the children.)

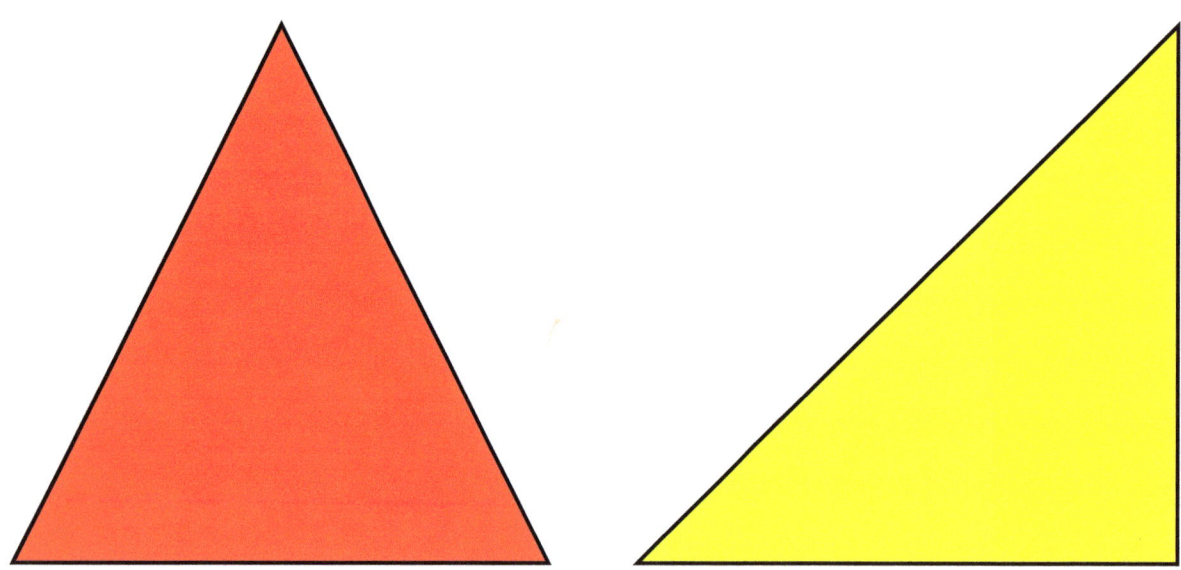

Point to a shape and ask, "What shape is this?"

Say: "Show me a _____." (Fill in the blank with the name of a shape.)

Polygons, Polyhedrons, and Three-Dimensional Shapes (Solid Figures) that Are Not Classified as Polyhedrons for Grades Kindergarten and Up

What is a polygon?

Instruction: A polygon has three characteristics.
1. It is flat, meaning that it is two-dimensional.

   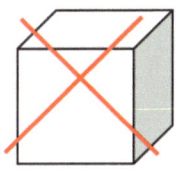   A 3-dimensional shape is not a polygon because it is not flat.

2. It is formed <u>entirely</u> of straight lines.

   A circle is not a polygon because it has no straight lines, and a heart is not a polygon because it has some curves.

3. It is closed—there are no open sides.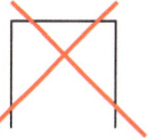

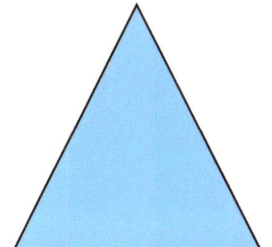 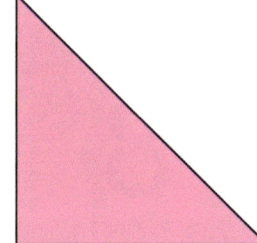  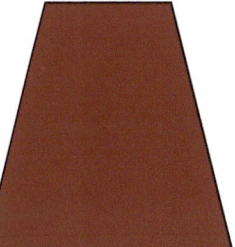

These are examples of polygons.

Instructions: Before viewing the next four pages, click the ear icon to hear "The Polygon Song" sung to the tune of "This Old Man Came Rolling Home." (Paperback readers can hear the tune by visiting the "Math Songs" page of the Habakkuk Educational Materials website at https://www.habakkuk.net/.)

Tri- means 3 △

quadri- means 4 ◻

penta- means 5 (and) 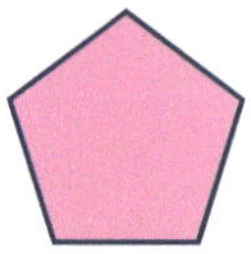 ⬠

hexa- means 6 ⬡

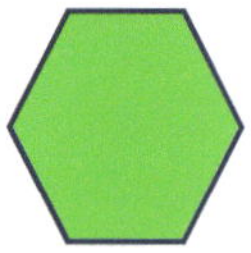

hepta- means 7 (and) 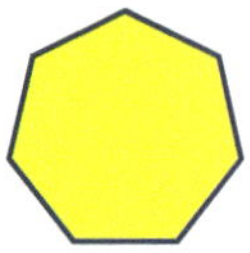 ⬣

octa- means 8 ⯃

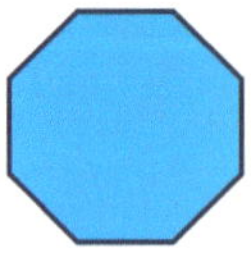

nona- means 9 (and) 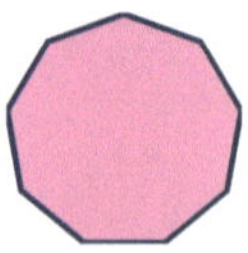 ⬣

deca- means 10 ⬣

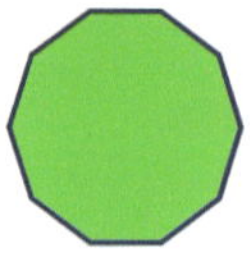

Polygons
(number of sides)

Instruction: You can use the illustrations in this box to explain to students what vertices and angles are. Younger students need only know that a flat shape has the same number of angles and vertices as it does sides.

A triangle has 3 vertices, one on each corner.

A square has 4 right angles.

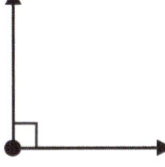

triangle

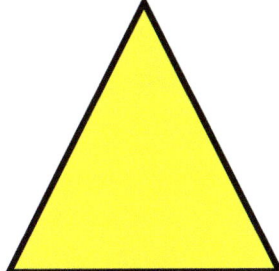

3

How many <u>sides</u> does a triangle have?
How many <u>angles</u>…?
How many <u>vertices</u>…?

(The answer to all three questions is three.)

quadrilateral

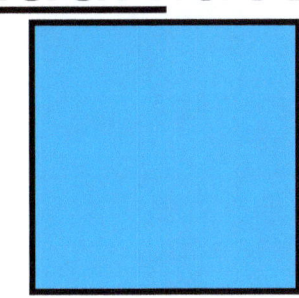

4

How many <u>sides</u> does a quadrilateral have?
How many <u>angles</u>…?
How many <u>vertices</u>…?

(The answer to all three questions is four.)

Examples of quadrilaterals include squares, rectangles, trapezoids, and rhombuses.

Polygons
(number of sides)

pentagon

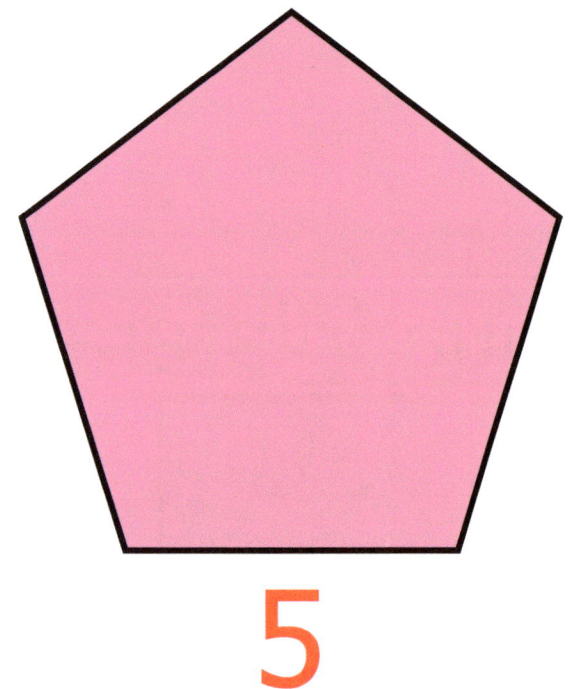

5

How many sides does a pentagon have?
How many angles...?
how many vertices...?

(The answer to all three questions is five.)

hexagon

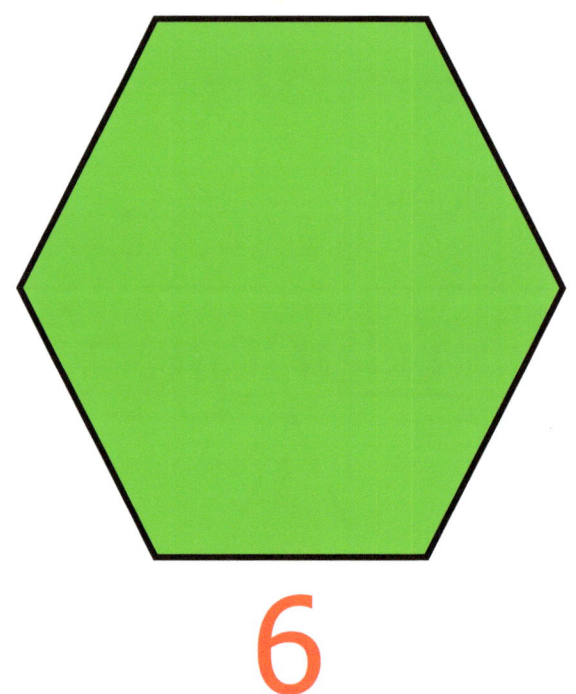

6

How many sides does a hexagon have?
how many angles...?
how many vertices...?

(The answer to all three questions is six.)

Polygons
(number of sides)

heptagon

octagon

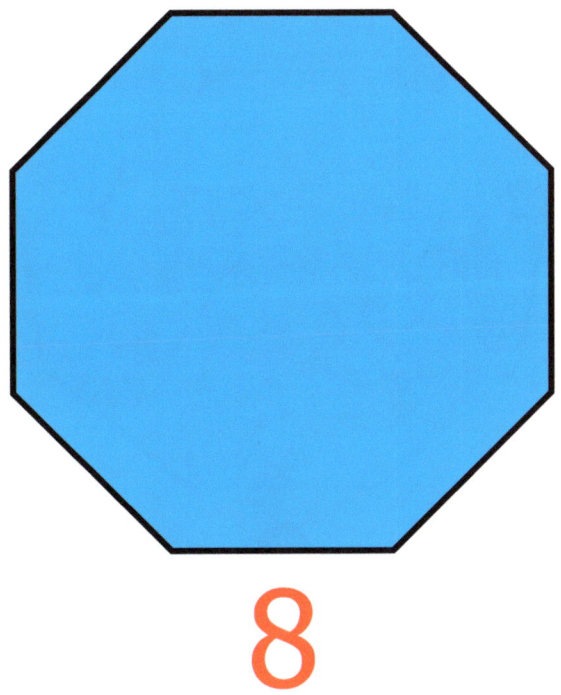

7   8

How many <u>sides</u> does a heptagon have?
How many <u>angles</u>...?
How many <u>vertices</u>...?

How many <u>sides</u> does an octagon have?
How many <u>angles</u>...?
How many <u>vertices</u>...?

(The answer to all three questions is seven.) (The answer to all three questions is eight.)

Polygons
(number of sides)

nonagon

9

How many <u>sides</u> does a nonagon have?
How many <u>angles</u>...?
How many <u>vertices</u>...?

(The answer to all three questions is nine.)

decagon

10

How many <u>sides</u> does a decagon have?
How many <u>angles</u>...?
How many <u>vertices</u>...?

(The answer to all three questions is 10.)

Instruction: Explain to students that an ***n*-gon** is a polygon whose number of sides is not known. A triangle can be referred to as a 3-gon. Can you tell me why? (Answer: It's a polygon with three sides.)

Polyhedron
(number of faces)

Instruction: A three-dimensional (3-D) figure that has polygons for faces is a polyhedron. *Poly-* means "many." In the pentahedron below, the faces are colored blue and purple.

tetrahedron | pentahedron

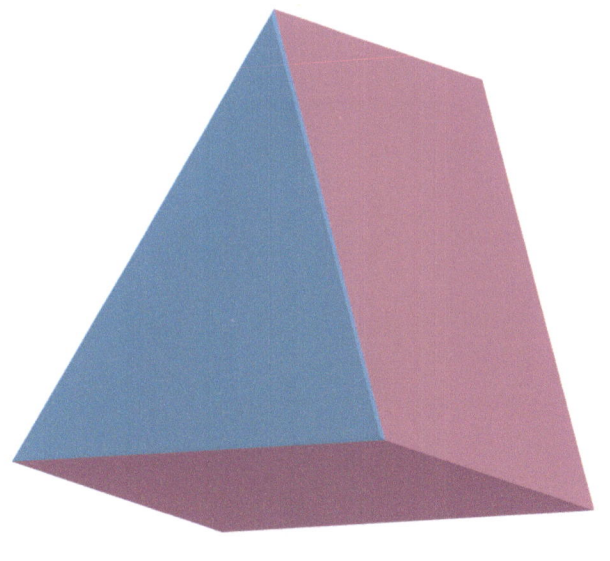

4

5

How many <u>faces</u> does a tetrahedron have?

How many <u>faces</u> does a pentahedron have?

_____

A *tetrahedron* has four faces, and a *pentahedron* has five.

Polyhedron
(number of faces)

Teacher instructions: Help students to understand the concept of faces by letting them count the six faces of a Rubik's Cube or something similar.

hexahedron (cube)

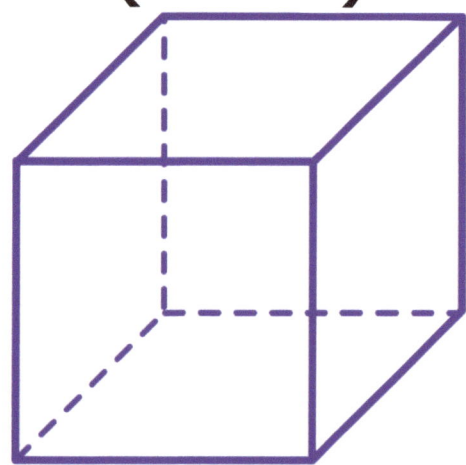

6

How many <u>faces</u> does a hexahedron have?

heptahedron

7

How many <u>faces</u> does a heptahedron have?

A *hexahedron* has six faces, and a *heptahedron* has seven.

Polyhedron
(number of faces)

octahedron

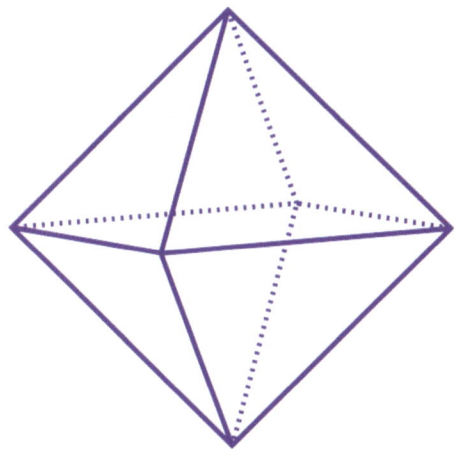

8

How many faces does an octahedron have?

_____

decahedron

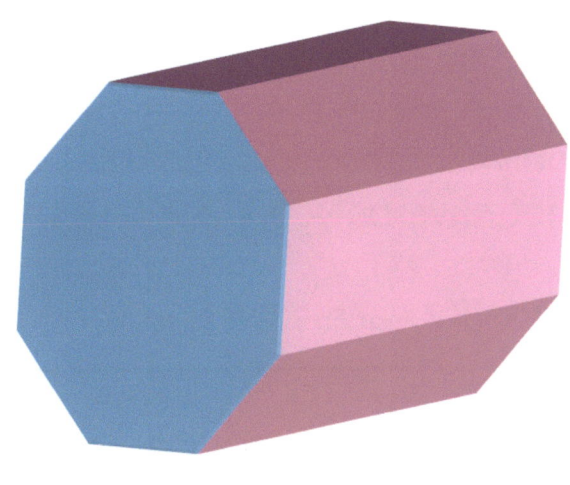

10

How many faces does a decahedron have?

An *octahedron* has eight faces, and a *decahedron* has ten.

Solid Figures that Are Not Classified as Polyhedrons

Instruction: A geometric solid (or three-dimensional figure) is not a flat shape. Even though the images on this page are solid figures, they are not polyhedrons like the other 3-D figures we just learned about. Do you know why? (The cone, cylinder, and sphere are not polyhedrons because they do not have polygons for faces.)

Identify each solid figure.

cylinder

A cylinder is shaped like a can.

cone

A cone is shaped like an ice-cream cone.

sphere

A sphere is shaped like a ball.

Name: _____

Matching Shapes with Shape Words

Directions: Write the name of each shape in the space provided. Circle all the quadrilaterals. The first shape has been identified for you.

Shape words: circle, cross, heart, hexagon, octagon, oval, ~~parallelogram~~, pentagon, rectangle, rhombus or diamond, square, star or decagon, trapezoid, triangle

▱ parallelogram

□ _____

✚ _____

◁ _____

⬠ _____

♡ _____

☆ _____

⬡ _____

○ _____

△ _____

▭ _____

⬭ _____

⯃ _____

▱ _____

Reproducible for non-commercial, classroom use only by Habakkuk Educational Materials

Geometry Test

Record the number of sides, angles, and vertices each polygon has:
3, 4, 5, 6, 7, 8, 9, 10, or *unknown*.

1. heptagon
2. quadrilateral
3. decagon
4. nonagon
5. *n*-gon

6. octagon
7. hexagon
8. triangle
9. pentagon
10. What is a 3-gon?

Identify the number of faces each polyhedron has:
4, 5, 6, 7, 8, or 10.

11. hexahedron
12. decahedron
13. heptahedron

14. octahedron
15. pentahedron
16. tetrahedron

17. Cross out the shapes that are not polygons.

  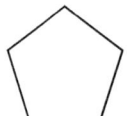 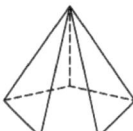 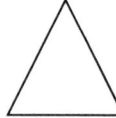 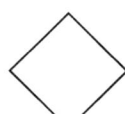  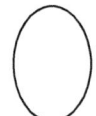

18. Circle all the quadrilaterals on the next page.

19. Which shape on the next page can be called a decagon?

Reproducible for non-commercial, classroom use only by Habakkuk Educational Materials

Directions: Write the name of each flat shape in the space provided. The first one has been done for you.
   circle, cross, diamond/rhombus, heart, heptagon, hexagon, nonagon, octagon, oval, ~~parallelogram~~, pentagon, rectangle, square, star, trapezoid, triangle

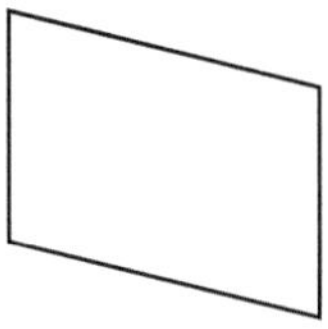

20. parallelogram 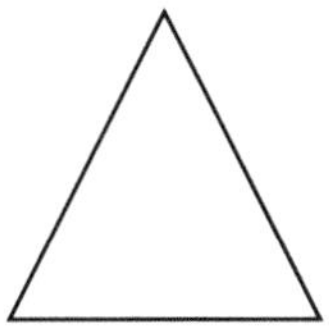   21.  _____   22.  _____   23. _____

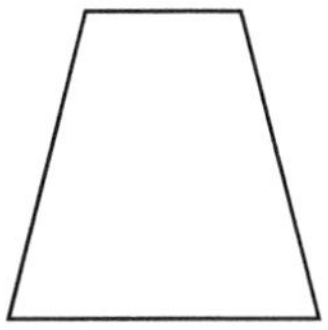

24. 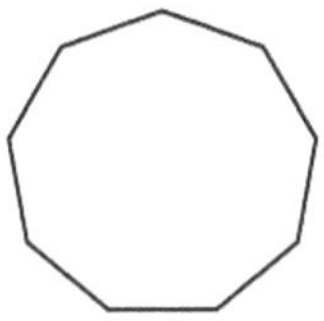 _____   25. _____   26. 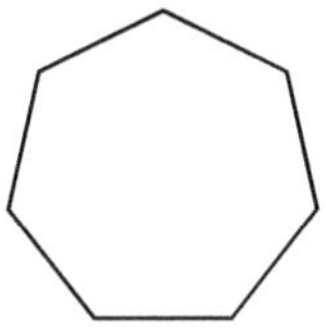 _____   27. _____

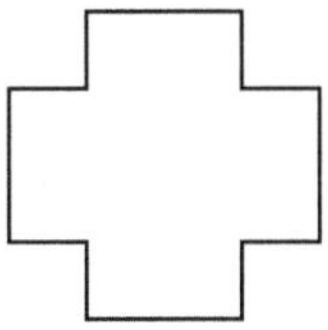

28. 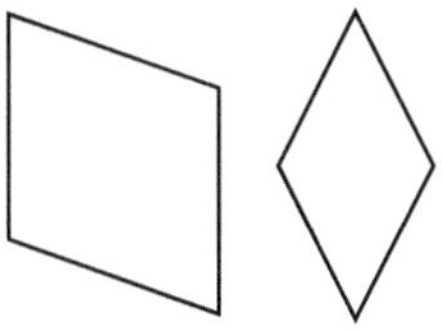 _____   29. 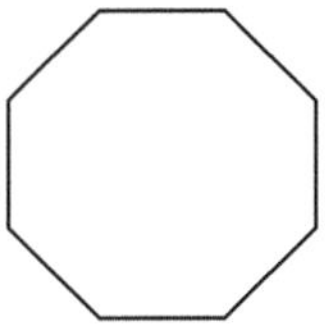 _____   30. 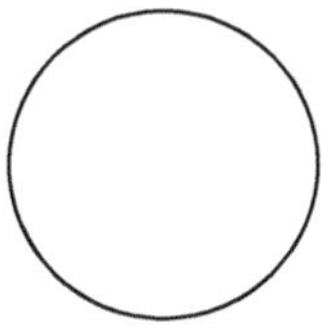 _____   31. _____

32. 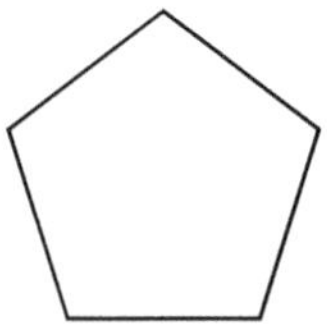 _____   33. 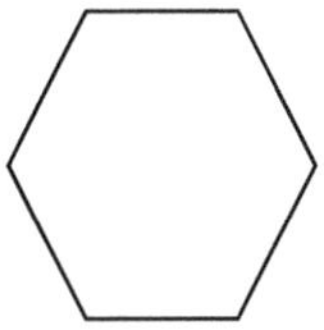 _____   34. 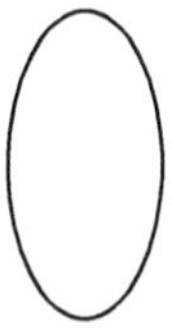 _____   35. _____

Directions: Write the name of each solid figure in the space provided.
   cone, cube, cylinder, sphere

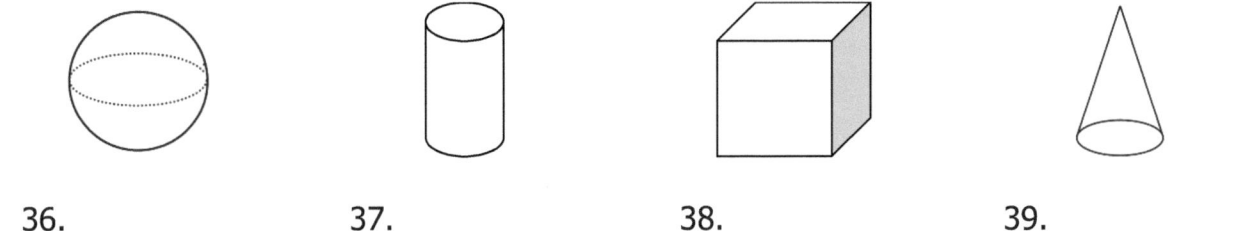

36. _____   37. _____   38. _____   39. _____

Reproducible for non-commercial, classroom use only by Habakkuk Educational Materials

Name: _____

Matching Shapes with Shape Words

Directions: Write the name of each shape in the space provided. Circle all the quadrilaterals. The first shape has been identified for you.

Shape words: circle, cross, heart, hexagon, octagon, oval, ~~parallelogram~~, pentagon, rectangle, rhombus or diamond, square, star or decagon, trapezoid, triangle

Shape	Name	Shape	Name
(parallelogram, circled)	parallelogram	hexagon	hexagon
(square, circled)	square	circle	circle
cross	cross	triangle	triangle
(trapezoid, circled)	trapezoid	(rectangle, circled)	rectangle
pentagon	pentagon	oval	oval
heart	heart	octagon	octagon
star	star, decagon	(rhombus, circled)	rhombus, diamond

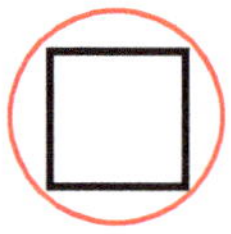

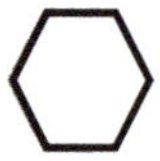

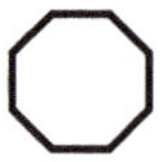

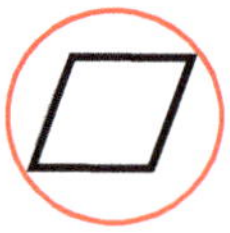

Geometry Test

Record the number of sides, angles, and vertices each polygon has:
3, 4, 5, 6, 7, 8, 9, 10, or *unknown*.

1.	heptagon	7	6.	octagon	8
2.	quadrilateral	4	7.	hexagon	6
3.	decagon	10	8.	triangle	3
4.	nonagon	9	9.	pentagon	5
5.	*n*-gon	unknown	10.	What is a 3-gon?	triangle

Identify the number of faces each polyhedron has:
4, 5, 6, 7, 8, or 10.

11.	hexahedron	6	14.	octahedron	8
12.	decahedron	10	15.	pentahedron	5
13.	heptahedron	7	16.	tetrahedron	4

17. Cross out the shapes that are not polygons.

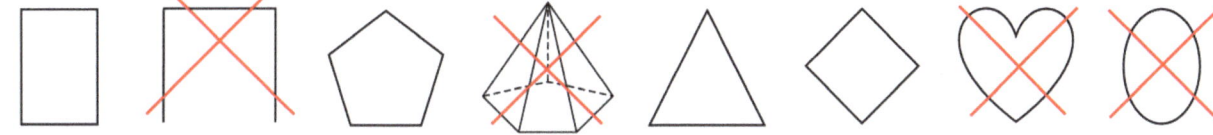

18. Circle all the quadrilaterals on the next page. The parallelogram, trapezoid, diamond/rhombus, rectangle, and square should be circled.

19. Which shape on the next page can be called a decagon? star

Directions: Write the name of each flat shape in the space provided. The first one has been done for you.

circle, cross, diamond/rhombus, heart, heptagon, hexagon, nonagon, octagon, oval, ~~parallelogram~~, pentagon, rectangle, square, star, trapezoid, triangle

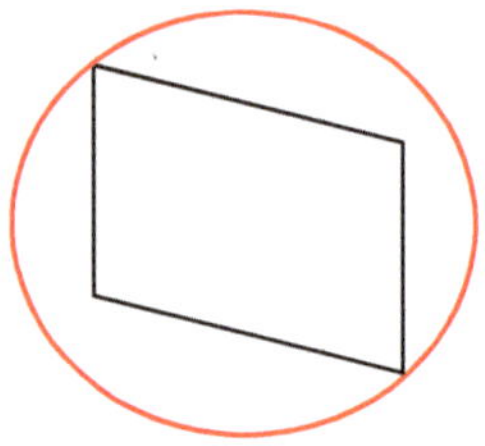

20. parallelogram 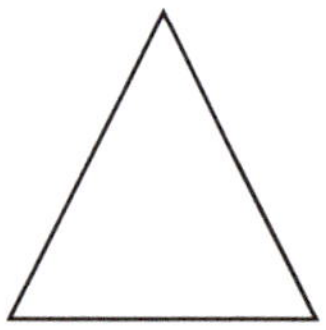   21. triangle    22. decagon / star    23. heart

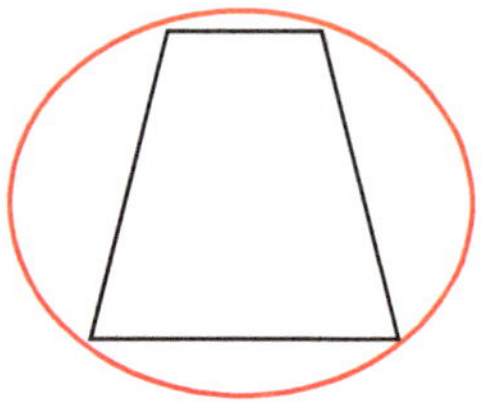

24. trapezoid 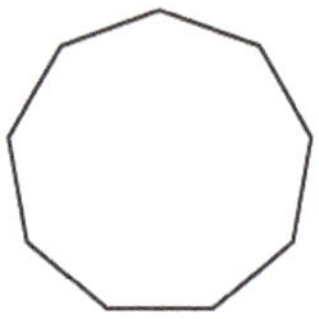   25. nonagon 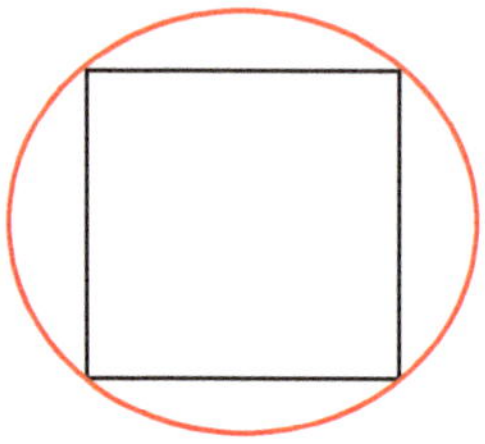   26. square 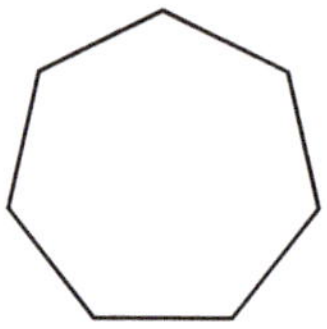   27. heptagon

28. cross 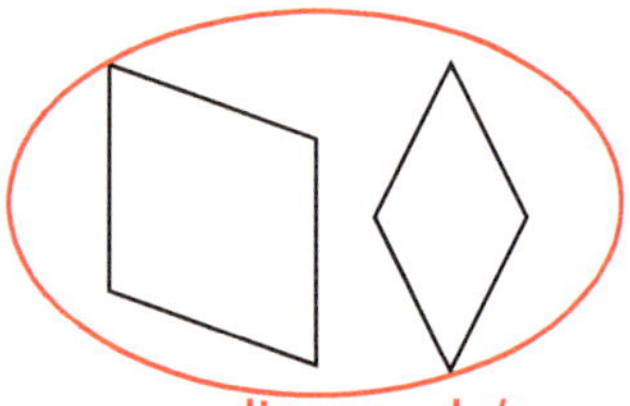   29. diamond / rhombus 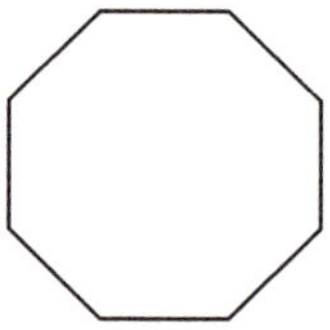   30. octagon 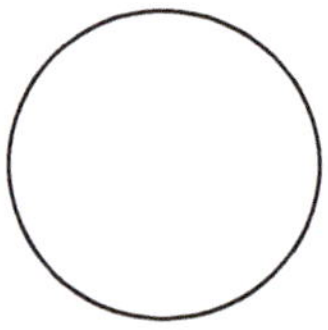   31. circle

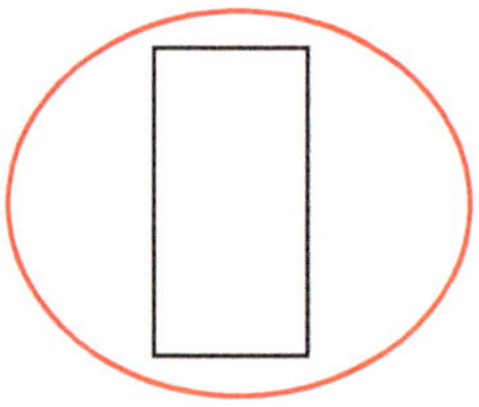

32. rectangle 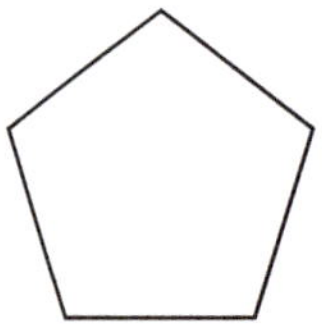   33. pentagon 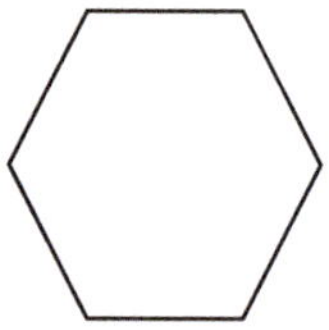   34. hexagon 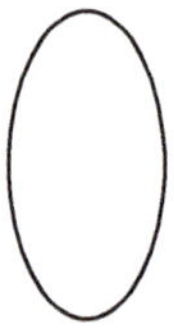   35. oval

Directions: Write the name of each solid figure in the space provided.
cone, cube, cylinder, sphere

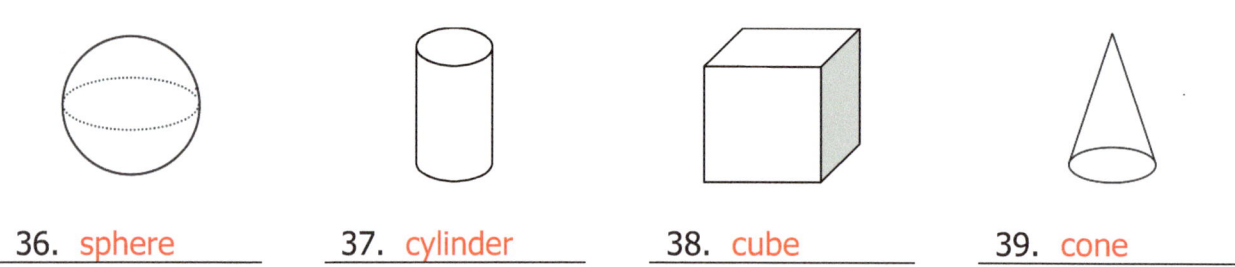

36. sphere   37. cylinder   38. cube   39. cone

www.ingramcontent.com/pod-product-compliance
Lightning Source LLC
Chambersburg PA
CBHW041538040426
42446CB00002B/135